1st Grade Science Volume 2

© 2013 Todd Deluca
OnBoard Academics, Inc
Newburyport, MA 01950

800-596-3175
www.onboardacademics.com

Table of Contents

Physical Characteristics of Animals

Can you match the animal with its outer covering?

  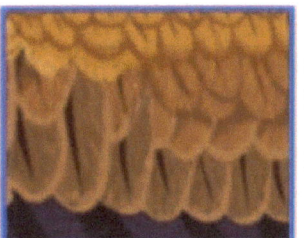 

These animals have very different physical characteristics. Physical characteristics describe what an animal looks like, including things like its shape, size, and color, and the type of body parts that it has. For example, birds have feathers, snakes have scales, deer have fur, and snails have hard shells.

Different Birds, Different Feathers.

See if you can match the picture and description of the feather type with the correct bird.

My feathers help me to stay camouflaged so I can hide.

My feathers are designed to be quiet when I fly. This helps me to hunt better.

I use my feathers to attract females of my species.

My feathers are short, broad and packed closely together to keep out water and keep in warmth.

Peacock

Penguin

Turkey

Owl

Different Animals, Different Teeth

See if you can match the picture and the description of the teeth with the proper animal.

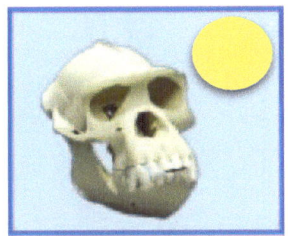 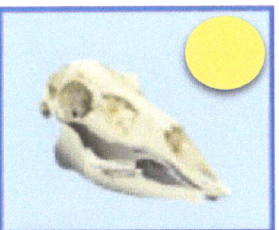 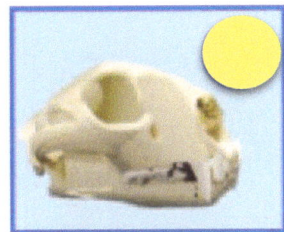 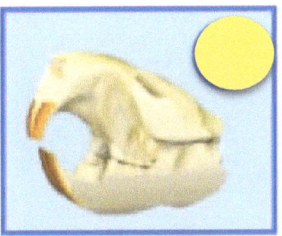

| I'm an omnivore. I use my incisors to bite into meat and my molars to grind up plants. | My molars are great for eating grass and other plants. I great at grinding plants. | My pointy canine teeth are great for hunting animals and tearing meat. | My huge incisor teeth are big and strong. I can even use them to cut trees. |

Cougar   **Chimpanzee**   **Beaver**   **Deer**

Different Animals, Different Feet

See if you can match the picture and the description of feet with the correct animal.

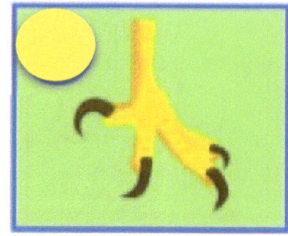

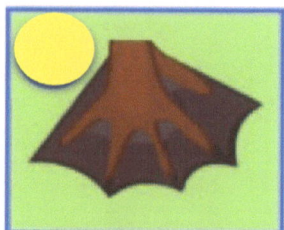

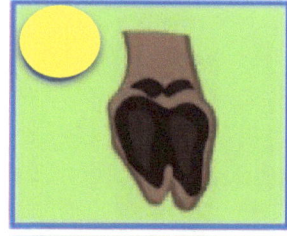

I'm an omnivore. I use my incisors to bite into meat and my molars to grind up plants.

My molars are great for eating grass and other plants. I great at grinding plants.

My pointy canine teeth are great for hunting animals and tearing meat.

My huge incisor teeth are big and strong. I can even use them to cut trees.

Squirrel

Platypus

Hawk

Mountain Goat

Sort the animals by the number of legs that they have.

0 legs	2 legs	4 legs

Starfish	Camel	Chicken	Whale	Grasshopper
Human	Snake	Butterfly	Goat	Hummingbird
Alligator	Spider	Octopus	Ant	Hedgehog

5 legs	6 legs	8 legs

Warm Blooded and Cold Blooded Animals.

Look at the following three illustrations. What happens to the dog and the lizard when the temperature (shown on the left) changes.

Dog: _____

Lizard: _____

The blood of warm-blooded animals stays at the same temperature, even when the outside temperature rises or falls. The blood of cold-blooded animals rises and falls with the outside temperature. Cold-blooded animals sometimes have cold blood, and sometimes have warm blood, but warm-blooded animals always have warm blood.

How do I breath?

All animals use either lungs, gills or spiracles to breath.

All animals need oxygen to breathe. Some animals use gills to get oxygen from the water. Some animals use lungs to get oxygen from air. Other animals have gills that turn into lungs. The smallest animals have spiracles: tiny holes on their bodies that let in oxygen.

Gills

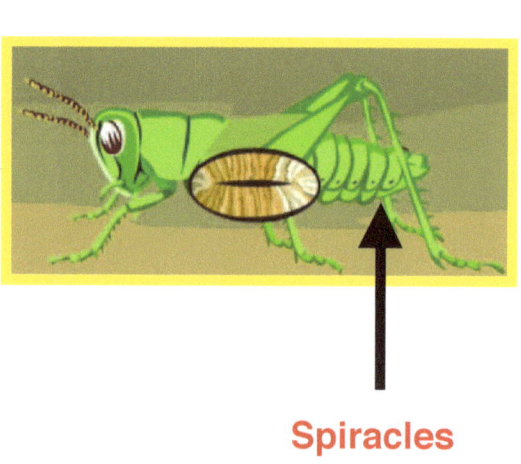

Spiracles

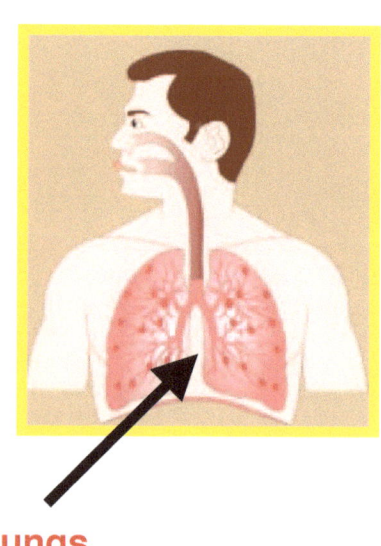

Lungs

How do I breath?
Label each animal **L** for Lungs, **G** for Gills or **S** for Spiracles

Mammals

Fish

Reptiles

Amphibians

Birds

Insects

Physical Characteristics of Animals Quiz

1. All animals share the same physical characteristics.
    True or false?

2. The feathers of _____ help them camouflage from predators.
    a. turkeys
    b. peacocks
    c. eagles

3. Which animal has wide molars to grind plants?
    a. cougar
    b. shark
    c. chimpanzee

4. Beavers have canine teeth for tearing flesh. True of false.

5. A _____ has strong sharp claws for digging tree bark while climbing.
    a. platypus
    b. hawk
    c. squirrel

Plant Parts

Do you know your plant parts?

See if you can match the picture of the plant part with the label.

flower

roots

seeds

leaf

stem

fruit

Let's learn about the different plant parts.

Study the illustration below.

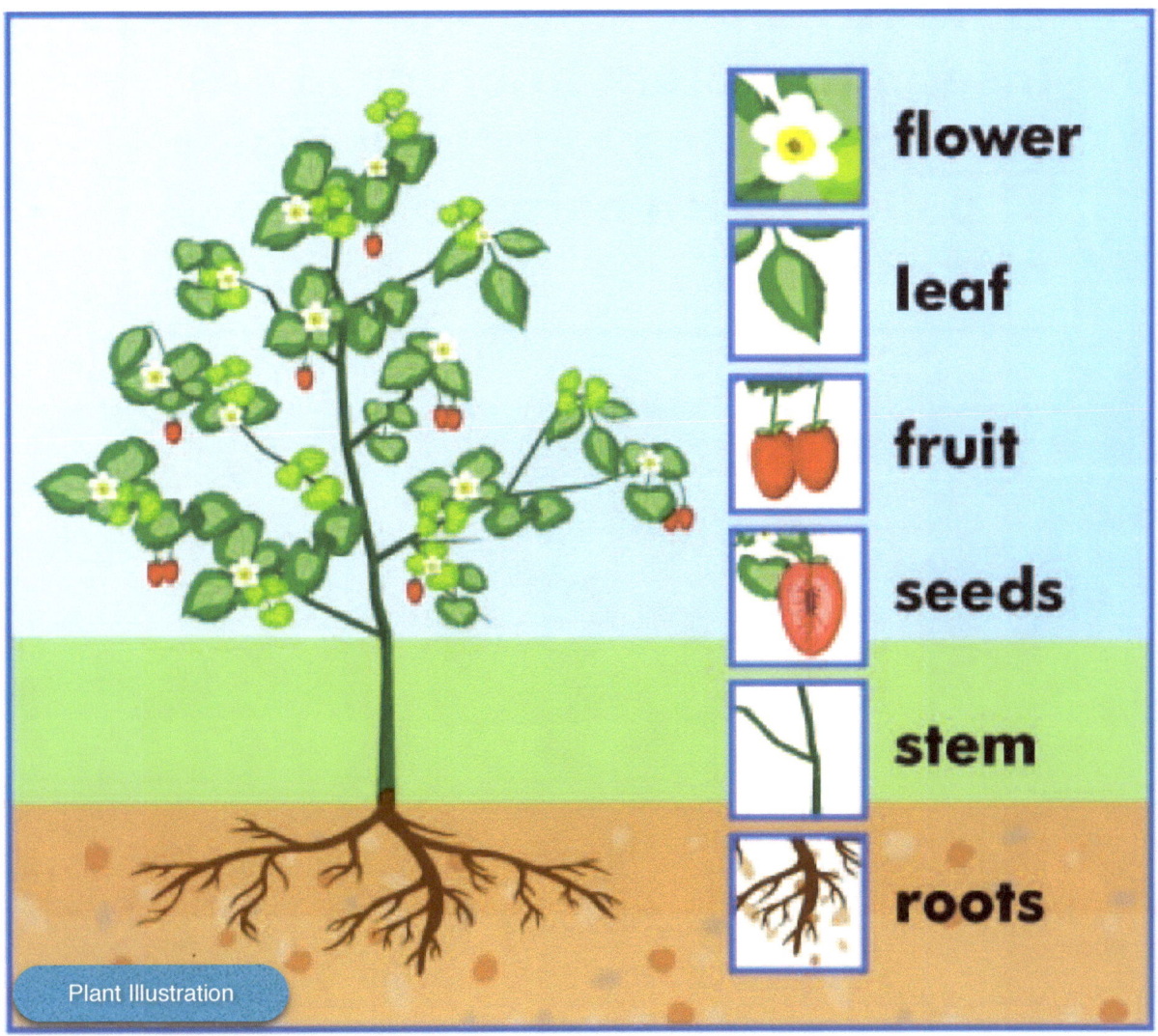

Plant Illustration

Flowers help to attract birds and insects to a plant in order to encourage **pollination**: the transfer of pollen which helps to create new seeds.

Leaves help to create food energy for the plant by turning water, sunlight and carbon dioxide into sugar and oxygen. Think of leaves as the plant's solar panels.

Fruit

Plants use fruit to find a new home for seeds since animals will eat the fruit, move off to a new location, and then eventually deposit the seeds on the ground in the new location. Since a fruit is a plant part that contains seeds, *vegetables* such as cucumbers, peppers, and pumpkins are actually fruit.

Seeds are the ways that plants reproduce. Plants put a lot of food energy into the seed so that the seed has enough energy to survive until it can produce its own food.

Stems carry water from the roots up to the leaves and carry sugars from the leaves to other parts of the plant. Stems also provide structural support for the plant.

Roots bring in water and nutrients from the soil. Roots also support the plant by anchoring it safely in the ground.

Match each plant part with its main function.

I do my best to attract pollinators.	I take water and nutrients from the soil.	If I germinate I will make a new plant.

Fruit   Stem   Seeds   Roots   Leaves   Flower

I transport water and food throughout the plant.	I help to bring seeds to new locations.	I use the sun, carbon dioxide and water to make food.

Which part of the plant are you eating?

Fruit   Stem   Seeds   Roots   Leaves   Flower

carrot

spinach

kidney beans

potato

tomato

cucumber

celery

asparagus

cauliflower

beet

corn

apple

pea

lettuce

broccoli

Plant Parts Quiz

1. Leaves help bring water an nutrients to the plant. True or false?

2. Which of the following plant parts provides support by anchoring the plant in the ground?
   a. stem
   b. seed
   c. root

3. The _____ carries water from the roots up to the leaves.
   a. seed
   b. stem

4. Sugar is transported from leaves to roots by the stem. True or false?

5. Oxygen is released by the leaves into the atmosphere. True or false.

6. Leaves can be compared to _____.
   a. wind turbines
   b. solar panels

Animal Adaptations; Teeth and Diet

Can you guess how many teeth humans have? Adults have more teeth than children.

16　　20　　32　　40　　64

　　www.onboardacademics.com

Can you guess how many teeth these animals have?

30   **300**   **42**   **0**

Different types of animals have different types and different numbers of teeth. Some animals, such as insects, birds, turtles, and some fish, don't have any teeth. What an animal eats (its diet) is an important factor in the number and type of teeth that it has.

How did you do?

 32

 20

 300

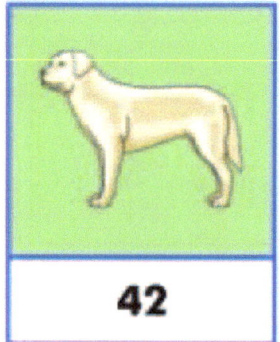 42

 30

 0

Types of Teeth

There are three types of teeth. **Incisors** are good for cutting and biting, **canines** are good for tearing, and **molars** are good for grinding.

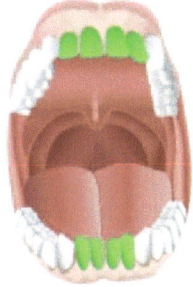

In a human mouth we have 8 incisors, two in the upper jaw and two in the lower jaw. These teeth have a flat thin edge that helps to bite and cut food.

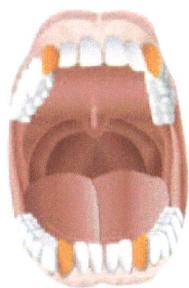

Next to the incisors are the canine teeth. These are long pointed teeth that help us to cut meat. There is a canine tooth on each side of the incisors on the top and bottom for a total of four canine teeth.

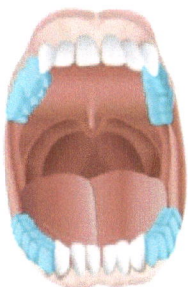

Behind the canine teeth are the molars. Molars are designed to help us to choose and grind food. Children have eight molars and adults have twenty.

Just like humans the type and number of teeth that an animal has is related to its diet. A tiger has very large canine teeth in its upper jaw to bite other animals.

A rabbit doesn't have any canine teeth.

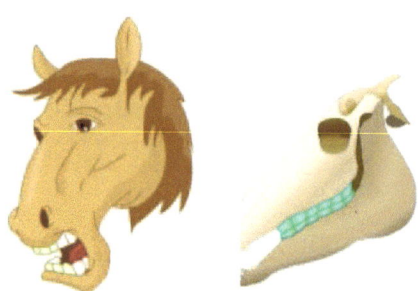

Horses and cows have incisors to help them to cut grass and other plants. They also have strong molars to help them to chew and grind rough vegetation.

Some animals like birds, fish and insects don't have any teeth at all.

Why do you think some animals do not have any teeth? _____

_____

_____

Identify each tooth and match it with its main job.

canines   molars   incisors

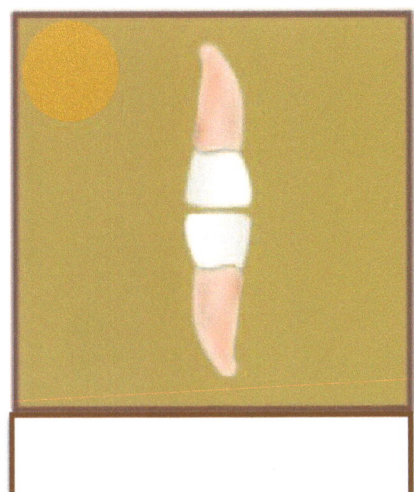 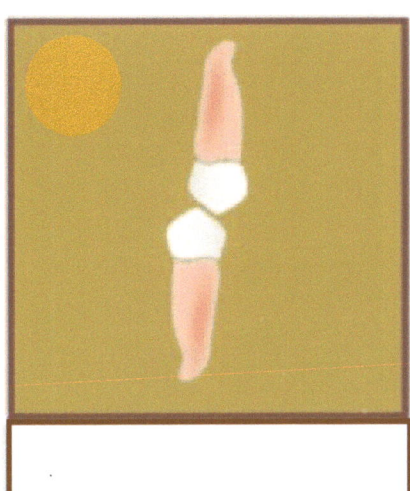 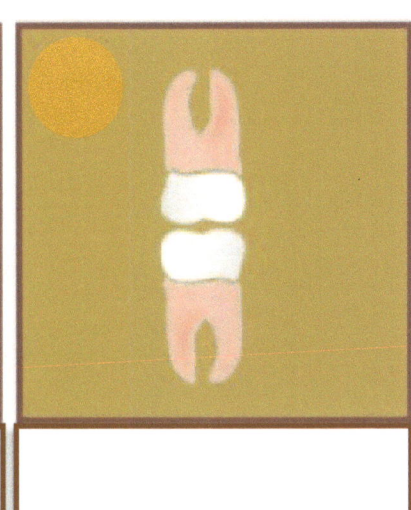

We have four of these teeth. We use them to tear things like meat.

We have eight of these teeth. They are great for cutting and biting.

We have 20 of these when we are older but only 8 when we are younger. We use these go grind and chew things.

Match these animals with their food.

Match these animals with their teeth.

What's for dinner?

Study the skeletons of teeth and look for the different types of teeth that exist in each skeleton.

Place a **P** for plant and or an **M** for meat in the boxes below each animal to indicate the type of food it eats based on what you know about its teeth.

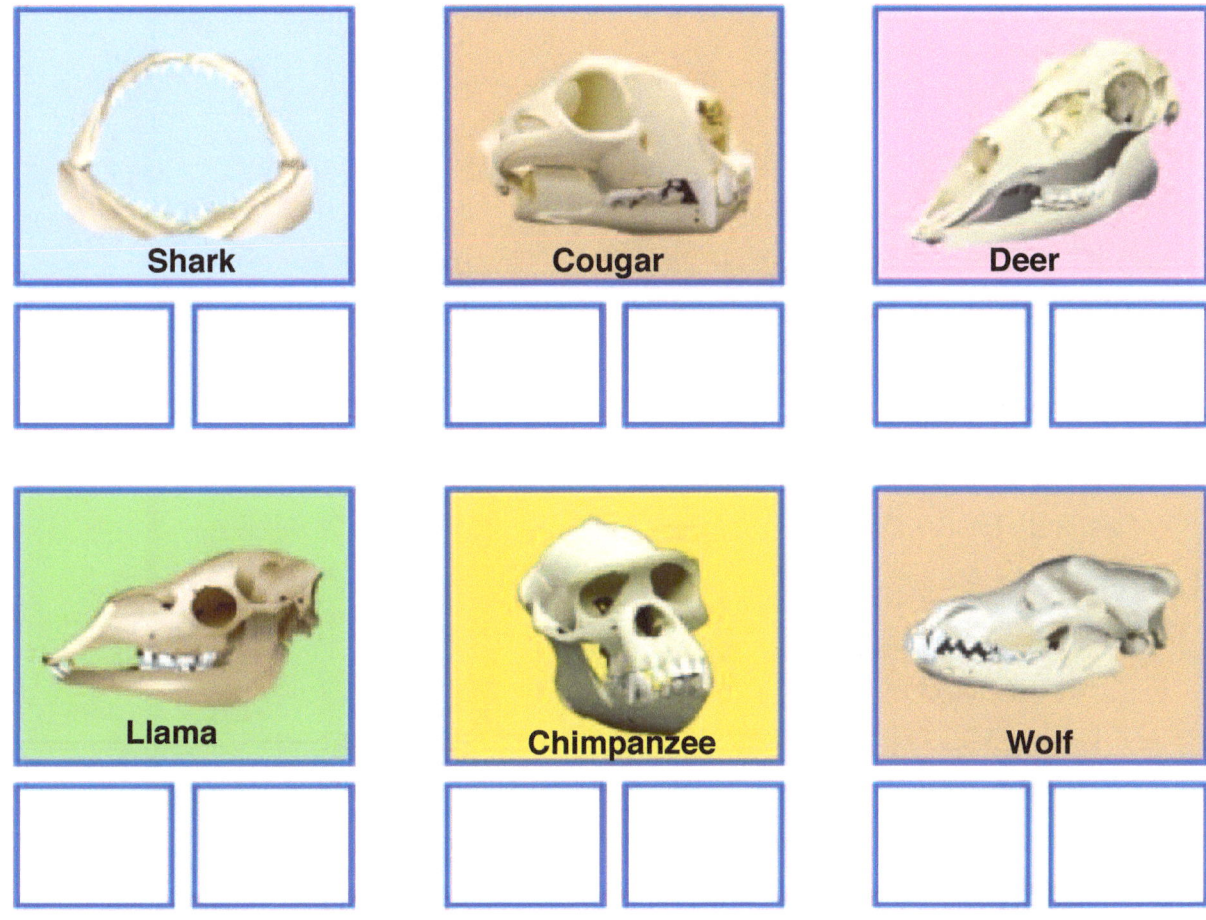

Animal Adaptation: Teeth and Diet Quiz

1. Insects have teeth. True or false?

2. All animals have the same number of teeth. True or false?

3. Teeth that help us to cut and bite food are called _____.
   - a. incisors
   - b. canines
   - c. molars

4. Children have _____ molars.
   - a. 5
   - b. 20
   - c. 8

5. Canine teeth help us tear meat. True or false?

6. Cows have _____ teeth to cut grass and other plants.
   - a. canine
   - b. incisor
   - c. molar

7. Meat eating animals have strong molars to cut meat. True or false?

8. Rabbits have sharp canine teeth. True or false?

Introduction to Classification

Classify these animals into two groups.

group one	group two

G-Giraffe   T-Tiger   S-Sea Turtle   B-Bear

P-Puppy   F-Fish   L-Lobster   D-Dolphin

To classify things means to put them into different groups according to a rule. One rule you might consider is "animals with legs and animals without legs" but there are many other rules that you may choose to sort or classify these animals. Select a rule other than the rule about legs or no legs and sort or classify these animals.

What was your rule? _____

_____

Classify these objects using the rule, "living and nonliving things".

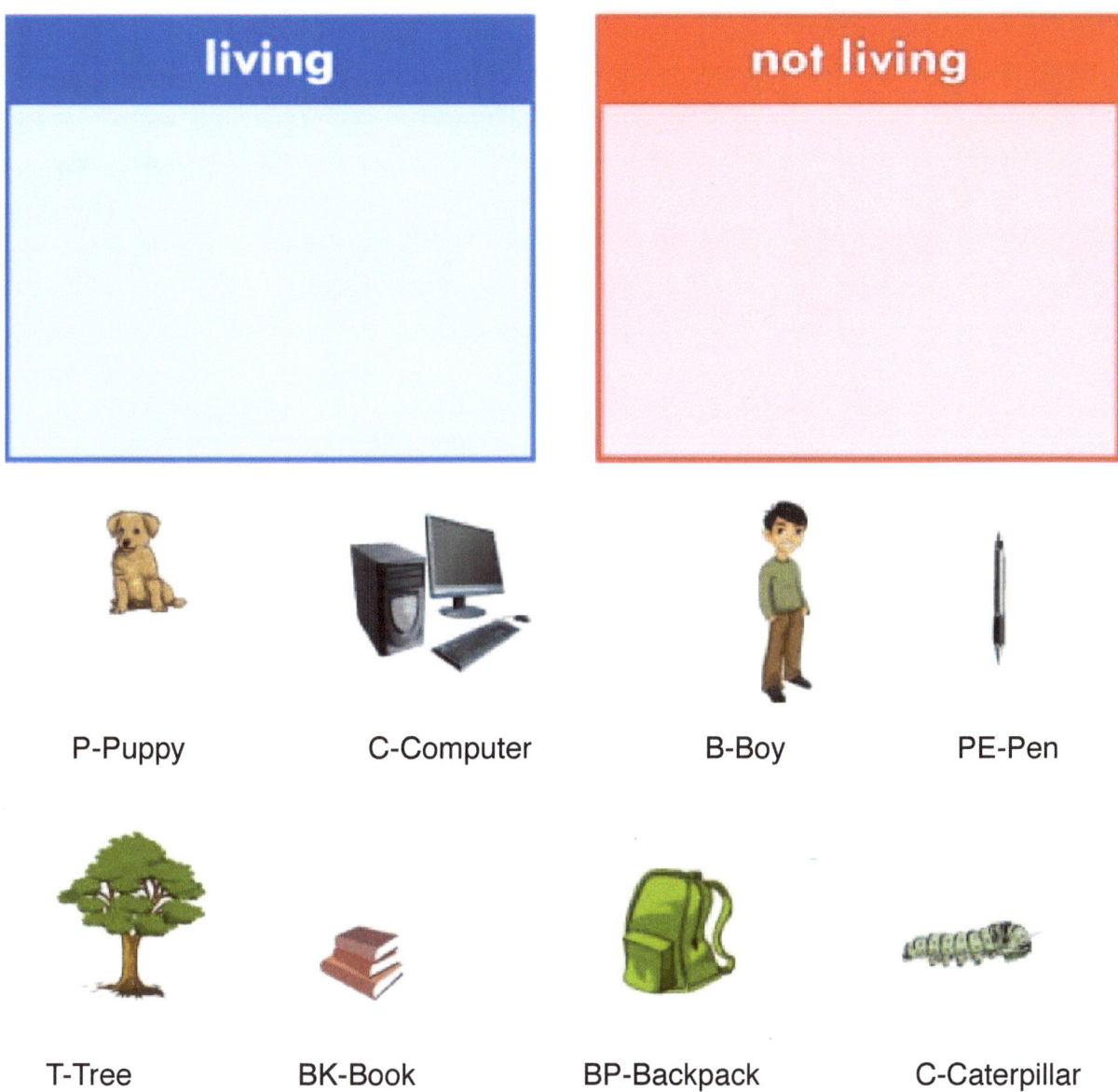

living	not living

P-Puppy   C-Computer   B-Boy   PE-Pen

T-Tree   BK-Book   BP-Backpack   C-Caterpillar

How have these animals been classified?

crustaceans   **reptiles**   **mammals**

insects   **birds**   **fish**

What rule has been used to classify these plants.
Write your answer in the boxes provided.

Sometimes we classify items using more than one rule.
Classify these objects using both rules.

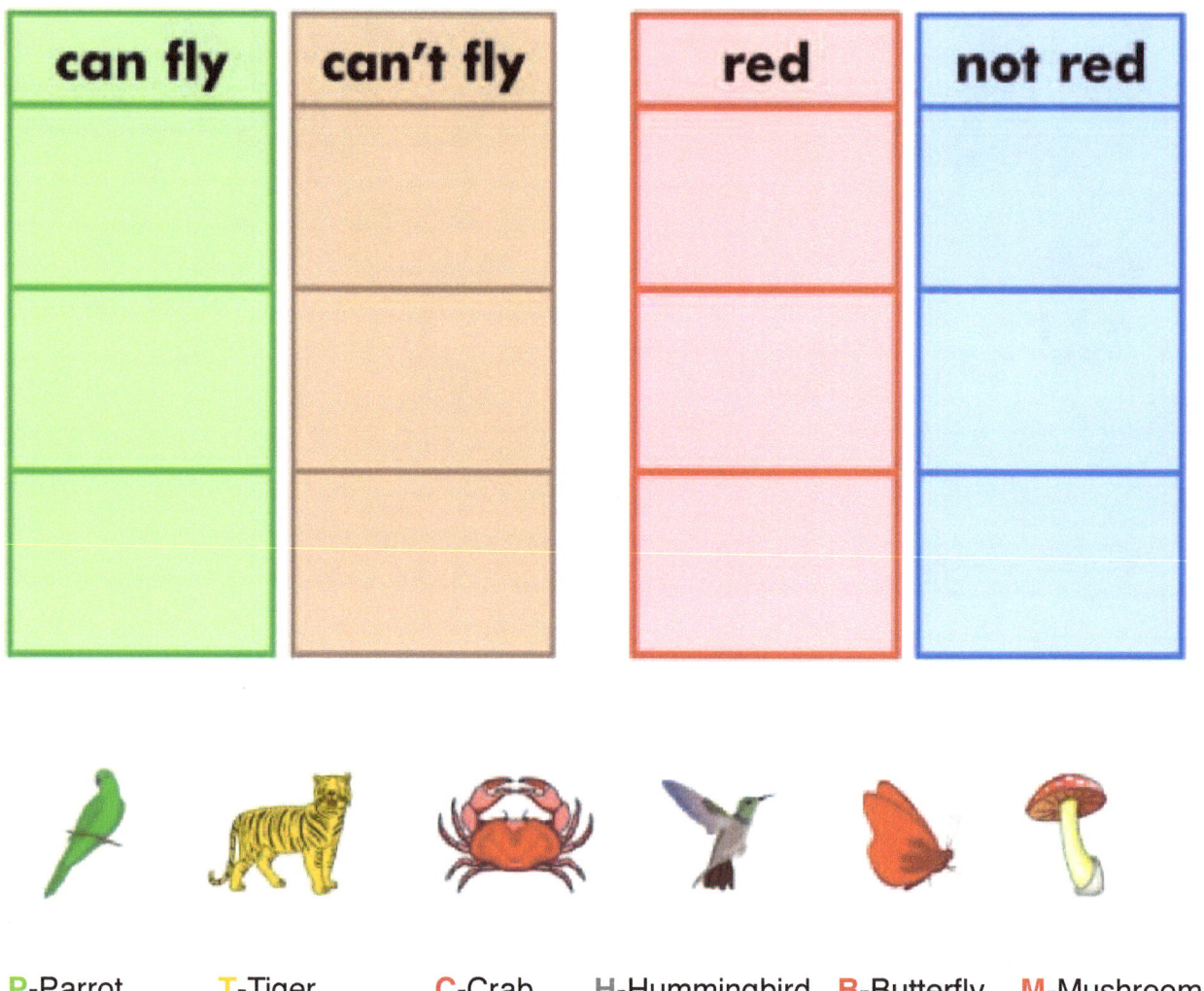

can fly	can't fly		red	not red

P-Parrot   T-Tiger   C-Crab   H-Hummingbird   B-Butterfly   M-Mushroom

We can classify into small groups when we use multiple rules.

Classify these items.

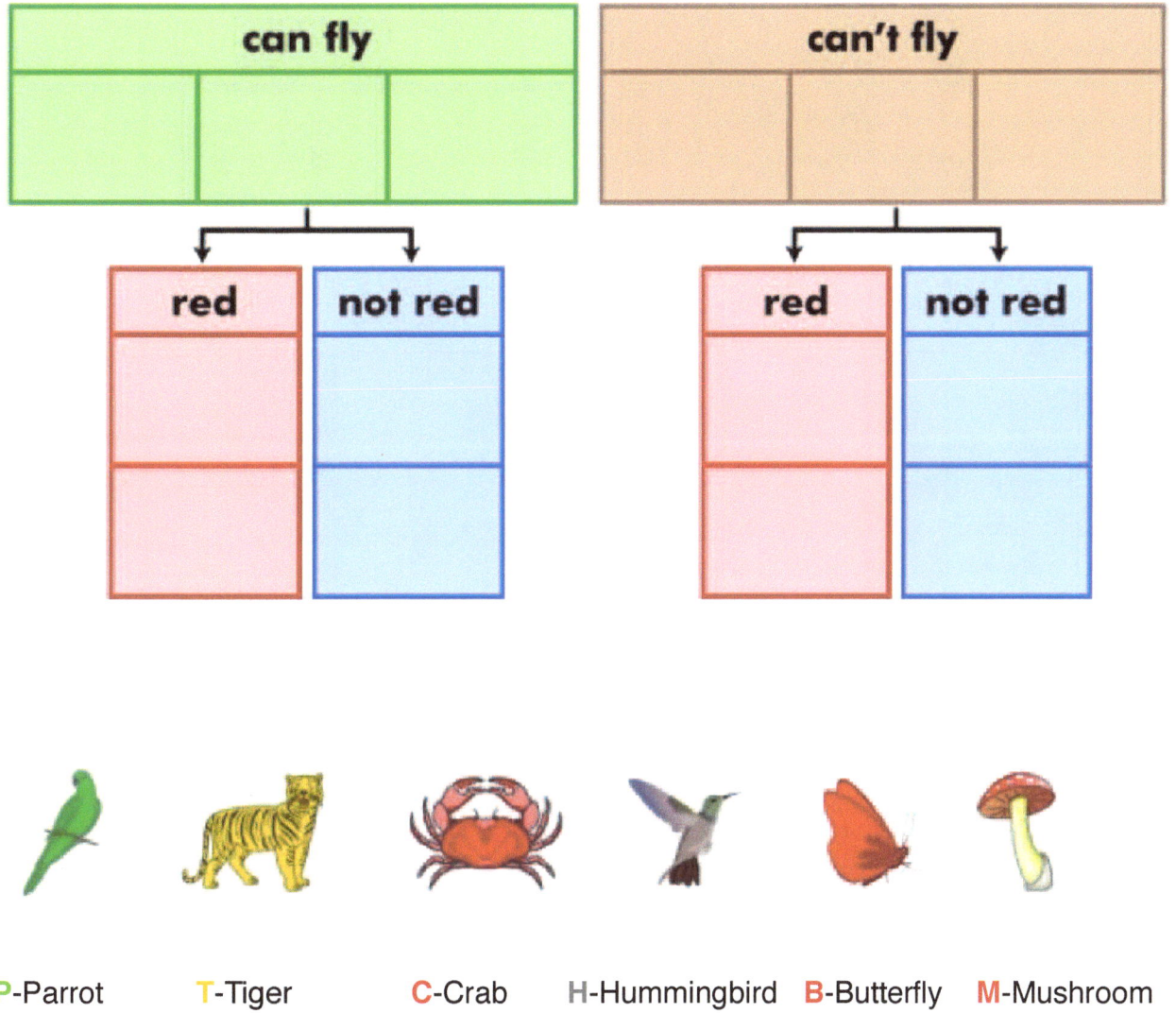

P-Parrot   T-Tiger   C-Crab   H-Hummingbird   B-Butterfly   M-Mushroom

Dichotomous Keys

Say *Dye Cot O'Mouse Keys*

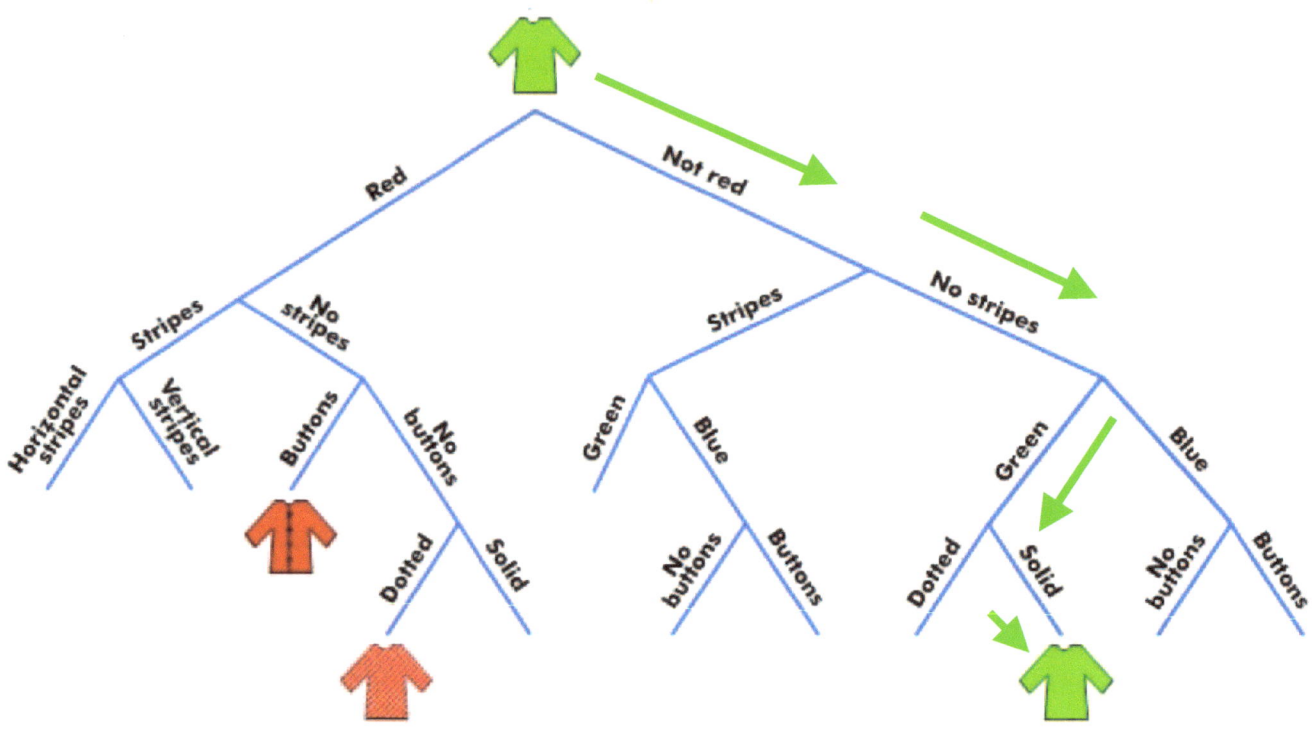

To use the dichotomous key, the item is classified by multiple rules. For this green shirt, we followed these rules.

Red or not red? Not red
Stripes or no stripes? No Stripes
Green or blue? Green
Dotted or solid? Solid

Are the red shirts placed properly? _____

Use the same method to classify these shirts. Draw the shirt in the proper position. If you don't have colors available write the name of the color under the drawing.

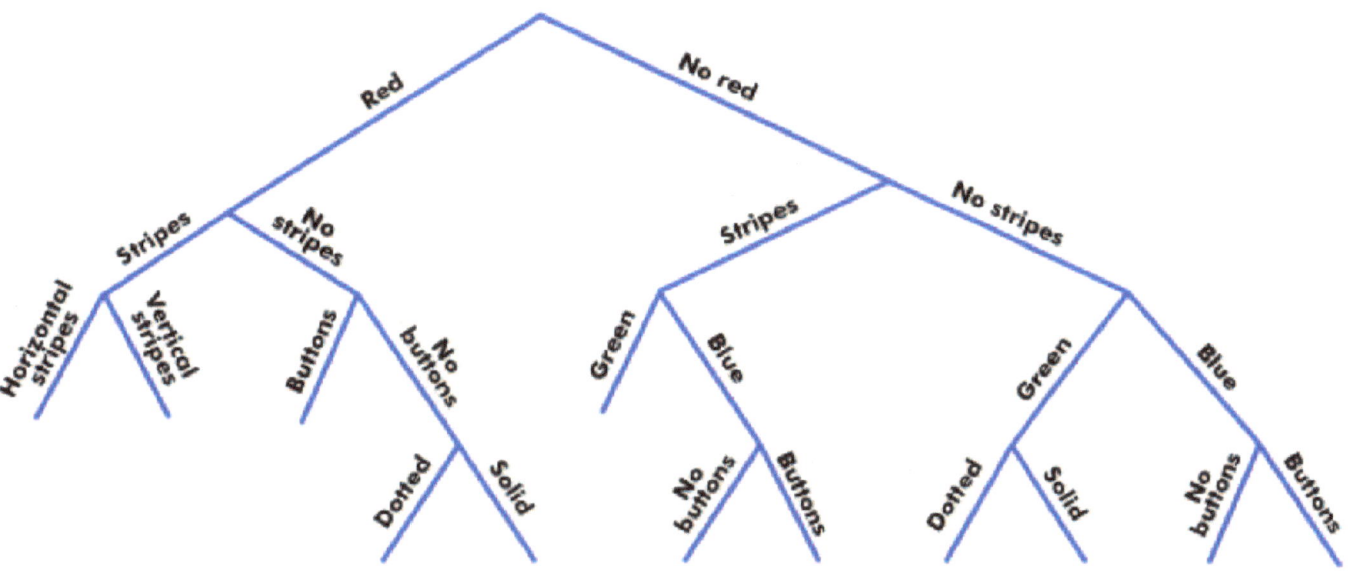

Introduction to Classification Quiz

1. Which of the following animals live in water?
    a. elephant
    b. whale
    c. deer

2. Which of the following animals live on land?
    a. fish
    b. shark
    c. snake

3. A plant is a living thing. True or false?

4. A car is a living thing. True or false?

5. An elephant is a(n) _____?
    a. mammal
    b. insect
    c. reptile

6. A fox is an insect. True or false?

7. Living things can be classified according to their color. True or false?

8. Scientists use a _____ to classify animals and plants/
    a. zigzag key
    b. branch key
    c. dichotomous key

Inherited Traits vs. Acquired Characteristics

What do babies know?

√ if babies know how to do it and X if they don't.

Inherited Traits and Acquired Characteristics

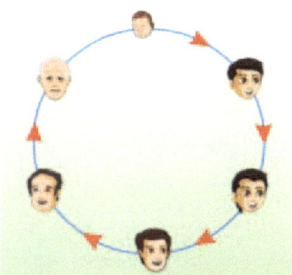

Throughout your entire life you are learning how to do new things and acquiring new skills.

For example you will learn how to walk, speak languages, drive a car and read and write.

Along the way you will get lots of help from parents, siblings and teachers as you learn how to do these new things and acquire new skills.

But who taught you how to breathe or sneeze or how to cry when you were a baby. No one did of course. You knew how to do these things when you were born.

We call things that you instinctively know how to do when you are born inherited traits. Things that you must learn to do are called acquired characteristics.

Acquired characteristics and inherited traits also describe a persons physical features, what you look like and your habits. You don't get to choose your eye color or you height and things like whether you have freckle, these are inherited traits.

But other things like a scar or a tattoo are acquired characteristics.

Sort these traits and characteristics.

inherited traits	acquired characteristics

Play soccer   **Blink**   **Smile**   **Cycle**   **Sneeze**   **Write**

Sort these physical characteristics.

A physical characteristic describes what your body looks like. Some physical traits are inherited (you are born with them), others are acquired (you get them as you get older).

inherited traits	acquired characteristics

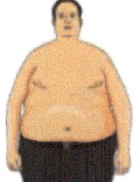     

Pot belly   Dimples   Tattoo   Freckles   Scar   Brown eyes

Are these characteristics inherited or acquired over time?

√ for inherited
X for acquired over time

best friend ☐

skin color ☐

language ☐

hair texture ☐

hair length ☐

favorite food ☐

Owen and Jack are identical twins. Which traits or attributes will they share, and which traits and attributes are likely to be different?

√ if they will be the same
X if they may be different

	Owen	Jack
hair color		
math scores		
shoe size		
height		
skating ability		
arm length		
favorite song		
eye color		
job		
number of kids		

Inherited Traits and Animals

Animals also have inherited traits. Label each animal with on of their inherited traits.

Climb

Hunt

Swim

Fly

Scientists call these traits adaptations. Animals have adaptations that help them to live in their environment.

Inherited Traits vs. Acquired characteristics Quiz

1. There are two types of traits: acquitted and inherited. True or false?

2. The traits that your parents give you are called:
   a. acquired
   b. inherited

3. Which of the following could be an example of an acquired trait?
   a. scar
   b. freckles
   c. eye color

4. An example of inherited trait would be _____.
   a. dimples
   b. scar

5. Your eye color and blood type are inherited traits. True or false?

6. Hair color is an acquired trait. True or false?

7. A striped pattern on a tiger's skin is _____.
   a. acquired trait
   b. inherited trait

Traits and Offspring

What do babies have in common with adults?
Sort the activities.

only babies	adults and babies	only adults

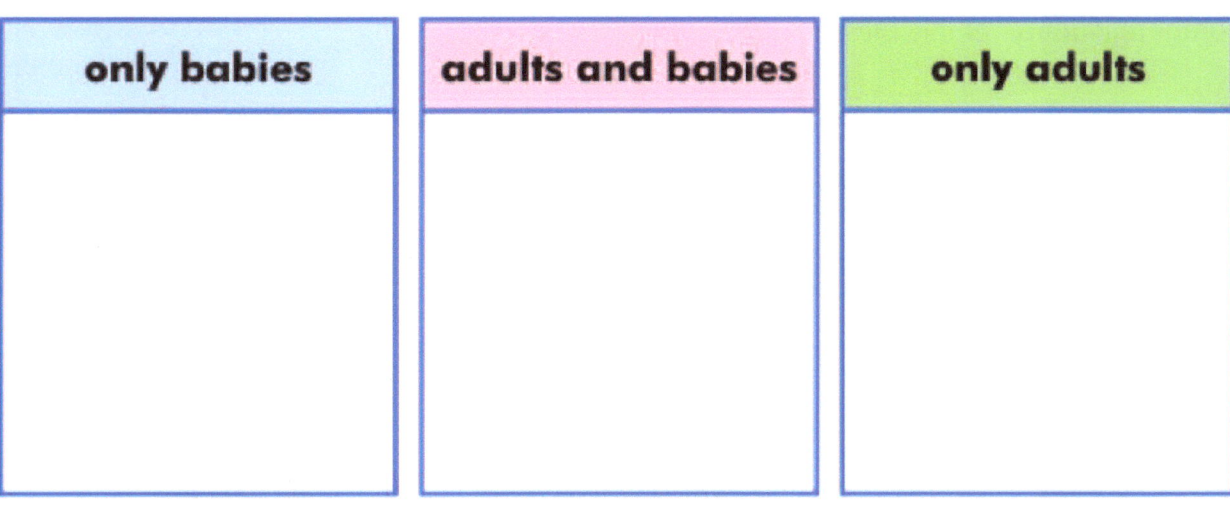

sleep   breathe   crawl   work

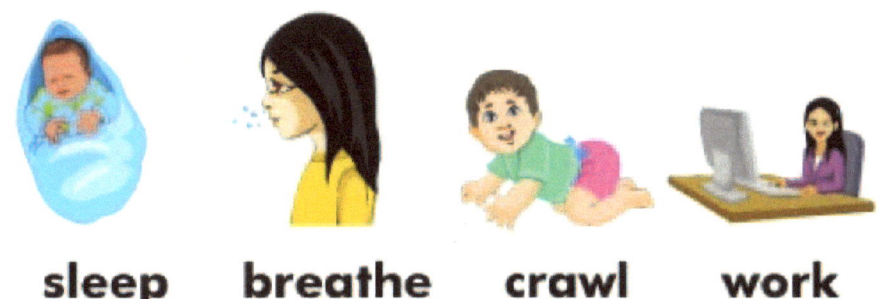

Babies and adults have some things in common, but they also have some things that make them different from each other. Let's explore similarities and differences in other animals.

Traits and Behaviors.

A trait is a term we use to describe how an animal looks. For example this bird had black feathers so we say black feathers is a trait of this bird.

Examples of some other traits are large teeth, yellow fur, webbed feet or a long tail.

A behavior is a term we us to describe how an animal acts. This bird might eat worms so we say that eating worms is a behavior of this bird.

Examples of other behaviors might be; a tiger likes to eat deer, a dog snores when it sleeps, a duck puts its head in the water to find food, a monkey climbed tree

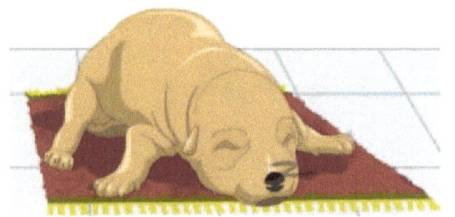

Remember, a trait describes what an animal might look like and a behavior describes how an animal acts.

Trait or Behavior?

Label **T** for trait or **B** for behavior.

This dog wags its tail when its happy.

The ears on this dog stand up.

This dog has brown fur.

This dog likes to bark.

This dog has four legs.

This dog has a black nose.

This dog likes to chase cats.

This dog never wags it tail when it's happy.

Lets compare an adult to an infant goose.

Study these three illustrations.

Place √ in either of the first two boxes to indicate if the description is a trait or a behavior. Then indicate if the characteristic is representative of either or both the adult and or infant with another √

	t	b	adult	infant
webbed feet				
black feathers				
flies				
swims				
black bill				
yellow feathers				

Compare the adult and infant rabbit in the same way you compared the geese.

	t	b	adult	infant
has fur				
has ears				
is blind				
drinks milk				
has four legs				
eats grass				

Let's compare an adult frog with an infant frog (tadpole) in the same was as we compared the rabbits and geese.

	t	b	adult	infant
has legs				
has a tail				
has a mouth				
has big eyes				
hops				
lives under water				
swims				
eats plants				
eats insects				

Traits and Offspring Quiz

1. A trait is an identifying feature of your personal nature. True or false?

2. Which one of the following is NOT a trait passed on from parents to child.
   a. skin color
   b. hair type
   c. coordination

3. The young inherit _____ from their parents.
   a. character
   b. traits
   c. behavior

4. Which of the following is NOT a behavior?
   a. a dog chasing its tail
   b. a dog having brown fur
   c. a dog wagging its tail

5. Which of the following describes behavior?
   a. broad nose
   b. detached earlobes
   c. getting angry easily

Newburyport, MA 01950

1-800-596-3175

OnBoard Academics employs teachers to make lessons for teachers! We create and publish a wide range of aligned lessons in math, science and ELA for use on most EdTech devices including whiteboard, tablets, computers and pdfs for printing.

All of our lessons are aligned to the common core, the Next Generation Science Standards and all state standards.

If you like our products please visit our website for information on individual lessons, teachers licenses, building licenses, district licenses and subscriptions.

Thank you for using OnBoard Academic products.